TRANSFORMING ANIMALS

# TURNING INTO A BUTTERFLY

by Tyler Gieseke

Cody Koala
An Imprint of Pop!
popbooksonline.com

**abdobooks.com**
Published by Pop!, a division of ABDO, PO Box 398166, Minneapolis, Minnesota 55439. 

Printed in the United States of America, North Mankato, Minnesota

102021
012022

THIS BOOK CONTAINS RECYCLED MATERIALS

Cover Photo: Shutterstock Images
Interior Photos: iStockphoto, 1; DebraLee Wiseberg / Getty Images, 5; Shutterstock Images, 6–17, 21; teptong / Getty Images, 18

Editor: Elizabeth Andrews
Series Designers: Laura Graphenteen, Victoria Bates

**Library of Congress Control Number: 2021942294**

**Publisher's Cataloging-in-Publication Data**
Names: Gieseke, Tyler, author.
Title: Turning into a butterfly / by Tyler Gieseke
Description: Minneapolis, Minnesota : Pop!, 2022 | Series: Transforming animals | Includes online resources and index.
Identifiers: ISBN 9781098241148 (lib. bdg.) | ISBN 9781098241841 (ebook)
Subjects: LCSH: Butterflies--Juvenile literature. | Insects--Juvenile literature. | Animal life cycles--Juvenile literature. | Butterflies--Metamorphosis--Juvenile literature. | Animal Behavior--Juvenile literature.
Classification: DDC 595.78--dc23

Hello! My name is

# Cody Koala

Pop open this book and you'll find QR codes like this one, loaded with information, so you can learn even more!

Scan this code* and others like it while you read, or visit the website below to make this book pop.

**popbooksonline.com/turn-butterfly**

*Scanning QR codes requires a web-enabled smart device with a QR code reader app and a camera.

# Table of Contents

Chapter 1

# Transforming Animals

Butterflies are pretty bugs that fly from flower to flower. They have six legs and two eyes. They have a **proboscis** for a mouth. It helps them suck **nectar** from flowers.

Watch a video here!

Butterflies are **transforming** animals. They grow through four steps.

The name for a group of butterflies is a flutter.

The steps are egg, caterpillar, chrysalis, and adult. The whole life cycle takes about one month.

Every butterfly begins life as an egg. The eggs stick to plants. Most **hatch** in three to five days.

Chapter 2

# Hungry Caterpillars

Caterpillars are as small as pin heads when they **hatch**. They will grow to be long and fat. They have several pairs of small feet. They wiggle to move.

Learn more here!

Some caterpillars have spikes on their backs. These protect them from **predators**.

Caterpillars eat a lot of leaves. They get bigger and shed their skin several times.

Chapter 3

# Time to Transform

After two weeks, it is time to **transform**! The caterpillar hangs from a plant. It creates a hard covering. This is a chrysalis.

Complete an activity here!

Inside the chrysalis, the caterpillar's body is turning into a butterfly. Two weeks pass. The chrysalis is now weak. It might be partly **transparent**.

Chapter 4

# Taking Flight

The adult butterfly breaks out of its chrysalis. It waits for its new wings to dry. Then, it can fly.

A butterfly's wings are made of thousands of tiny scales.

Learn more here!

A butterfly's brightly colored wings warn **predators** that the wings are toxic. Some butterfly wings have spots that look like eyes. These can help the butterfly find a **mate**.

Butterflies are adults for two to six weeks. They drink **nectar**. They make new eggs with a mate. The life cycle starts again! Butterflies are amazing **transforming** animals.

# Life Cycle of a Butterfly

# Making Connections

## Text-to-Self

Which is your favorite step in the butterfly life cycle? Why?

## Text-to-Text

Have you read other books about bugs? How are those bugs similar to butterflies? How are they different?

## Text-to-World

What is another transforming animal you know about? How is its life cycle similar to or different from a butterfly's?

# Glossary

**hatch** – to break out of an egg.

**mate** – a partner animal of the same kind. Together they make new eggs or babies.

**nectar** – a sweet liquid inside flowers.

**predator** – an animal that hunts other animals for food.

**proboscis** – a body part like a tube that unrolls and sucks up food.

**transform** – to change into a new shape.

**transparent** – clear or see-through.

# Index



Online Resources

**popbooksonline.com**

Thanks for reading this Cody Koala book!

Scan this code* and others like it in this book, or visit the website below to make this book pop!

**popbooksonline.com/turn-butterfly**

*Scanning QR codes requires a web-enabled smart device with a QR code reader app and a camera.